Heroes' Hearts

The Legend & Legacy of the
Only Sweetheart Souvenir from the
Battle of Pearl Harbor

By
Gordon Richiusa

HEROES' HEARTS ®

By Gordon Richiusa

Copyright © 2016
ISBN: 978-0-9829926-2-3

All rights reserved. No part of this publication may be reproduced, stored in a retrieval system, or transmitted in any form or by any means, electronic, mechanical, recording or otherwise, without the prior written permission of the author.

Published in the United States by
Five Birds Publications,
Laguna Woods, CA.

HEROES' HEARTS ®

By Gordon Richiusa

Acknowledgements

This book is not intended as libelous, slanderous or to caste a negative pall on any person either living or dead, or branch of service, culture, or ethnicity. Most of this book is derived from a first person memoir and transcript of a recorded interview from an interview with a Pearl Harbor Survivor Marine, Salvatore Richiusa, which took place in 2001. Any opinions or statements made herein, therefore are the sole responsibility of the Salvatore Richiusa, who died in August of 2015. In addition, I would like to acknowledge the help of several people in uncovering the facts (which are still be uncovered at the time of this first printing) surrounding my father's statements, or in the creation of the exact replica Heroes Hearts® bracelet, in the manufacturing process, in the creation of the non-profit, or in the support projects which now are underway and support our mission to help heroes, one good deed at a time. Those who should be acknowledge include (in no particular order) John Lombardo, Ed Hoffman, Daniel Martinez, Gary Richiusa, Judy Nichols, Lynda Lee, Michelle Manu, Glen Stevenson, Frank Wimberly, Joane Shamma, Mark Gottlieb, Irene Hirano Inouye, and Dana Stamos.

Cover and interior photos by Photo Editor: Lynda Lee of L&LMagazine.
Cover design: Dana Stamos, Managing Editor of L&L Magazine.

HEROES' HEARTS ®

By Gordon Richiusa

DEDICATION

To all daughters, mothers, sons, fathers, of all cultures and every corner of the globe, in the spirit of aloha, in memory of Salvatore Richiusa and the love of his life, his wife of 72 years, my mother Flora "Mae" Villani.

--Gordon Richiusa

HEROES' HEARTS ®

By Gordon Richiusa

Index

Part One: The Legend "Two Hearts As One"………….p. 11

Part Two: Sam's Song/Transcript of Interview …….p. 21

Part Three: Photo Gallery……………………………………...p. 49

Part Four: The Legacy—Heroes' Hearts Inc …………p. 81

HEROES' HEARTS ®

By Gordon Richiusa

Part One

The Legend and Legacy of Mae and Sam
Two Ordinary Americans

The true beauty of my father and mother, and the story contained herein, is that they were just ordinary Americans. They had flaws, problems, made mistakes, but when the time came to prioritize their most important, life and death values, they both met the challenges with style and courage. These qualities of unquestioned loyalty, and a willingness to sacrifice everything for the others, as we know is the common thread for those who were part of what has become to be called *The Greatest Generation*. It is also a requirement for every good parent and person of service.

As a journalist who has interviewed hundreds of people over the years, I can say with certainty that every human has a fascinating story to tell. It is perhaps my failing that I did not know the importance of my parents' story while they were alive. I will do my best, now that

I've uncovered the facts to some degree, to do their collective story justice.

My father was a Pearl Harbor Survivor, a Marine (until the day he died, like all Marines). However, until the end of his life, these facts were incidental to the things he held most dear: His wife, family, and core ethical values, not the least of which was demonstrated in the desire to be his own best example. He was a proud American, a life long Democrat, a union man, a lover of science, facts, and baseball, dyslexic, an inventor with a wry sense of humor, one of the first heart bypass patients in the U.S., one of eight children of two Sicilian Immigrants, and a pretty great father to five children. My grandfather, Frank Richiusa (My middle name is Frank and my surname means literally, "Re-Close") was destined to be a priest, but when a visiting cardinal ordered my grandfather to pour unused holy oil down the drain instead of giving it to the poor (as was his habit apparently), my grandfather protested by quitting the priesthood.

My mother was a typical Italian mother who had been at least verbally abused by an over bearing father and even more typical Italian mother (also both

By Gordon Richiusa

immigrants);one of four children my mother overcame obstacles from the very first day of her life as she was thought to be dead at birth, wrapped in blankets and left on the floor behind a couch (some versions say "behind a stove" but it doesn't really matter). Small and malformed, she was born without a hip on the right side and only because her grandmother could hear muffled sounds coming from inside the blankets did my mother ever make it to the second day of life. So, it was something of a miracle that my father and mother were even born, let alone ever met and got married.

Sadly, I did not know the significance to the story of my parent's life until after they had both passed. Luckily, like most parents, they made every sacrifice to insure that I'd make good decisions in my life, paying for my schooling, and setting a good example for me to become the person I am today. When I made a mistake, my father would say something like, "Serves you right" or "That's not how I would have done it."

One of the significant details of my life is that I am a journalist, and in 2002 I was allowed to travel to Oahu for the 60th Anniversary of Pearl Harbor. I really didn't have any idea what that meant, because my father (again, like most veterans) never talked about his service. I remember going to my dad and telling him I was going to publish a story in the Los Angeles Daily news about him and his story. He seemed mildly pleased, so I asked if it would be alright to video tape an interview...to make sure what I wrote would be accurate. He agreed and so my journey began. What he told me was so incredible, that I just had to share it. The interview is transcribed and makes up the bulk of this book.

For now, let me say that I was very impressed, but even after the interview I didn't really understand the full impact and importance of his story. At Pearl Harbor (you will read his own words shortly) my dad was a Sicilian-American engineer who had enlisted in the Marines and whose crew built the barracks at Hickam Field. According to my father's recorded testimony, just like in the movie Pearl Harbor, when the shooting started, his crew had to break into the armory and get ammunition as well as weapons. As engineers, they were not issued any combat weapons. So, they mounted guns on their trucks, skip loaders, and tractors. After the shooting stopped, the Marines were a big part of the mop up. Some planes had crashed on land and they had to check to see if there were any survivors.

The first plane they came across was near a pineapple field. My dad said that a "native" came to the truck they were in and handed a knife, which the native said he'd killed the pilot with, claiming that the knife had been taken from the still living pilot.

My dad indicated that he did not look into any of the details of the story, but that he took the knife and passed it along later to one of my cousins. At another downed Japanese plane, two more "trophies" were taken, two pieces of aluminum from the cockpit of a plane where the pilot was already dead.

One piece (a small red one) was placed inside my dad's logbook and forgotten until two years ago (more on that later). Another larger piece he made into a bracelet for his wife-to-be, my mother Flora Mae Villani.

It was a beautiful design, what I call, "Two Hearts Beating as One," a linked hearts symbol that represents to me how all heroes' feel about their chosen path. Anyway, I knew that the bracelet was real, because my mother had been wearing it as long as I could remember. My parents had never told me where it had come from until that interview.

My father came home after Pearl Harbor for a short time, married my mother, gave her the bracelet and was returned to fight in some of the bloodiest battles of the Pacific Campaign.

I ended my video interview, eyes wide and mouth agape, with the question, "Do you have ANY good memory, whatsoever from your experiences of World War II?" My father answered quickly, "Getting married was a good thing that happened, but my best memory was coming home and meeting my two and a half year old daughter for the first time. My father did not know that my mother was pregnant with my oldest sister Lynda until she was born. She had been conceived during his short visit home after my parents' elopement.

Fast-forward 72 years later to the memorial after my mother died. It was an emotional ceremony, and later that day my father asked his five children (in order of appearance Lynda, Judy, Gary, Gordon and Cheryl) to take my mother's possessions and divide them between us. I didn't really want anything, but at second thought I asked, "Can I have the bracelet?"

My father looked at me and smiled, "I think you *should* have it," he said. "Your mother would like that, but I want to give it to you because I now you're going to *do something good with it."*

Naturally I said, "Of course I will," not realizing what that was going to mean.

Out of character to his usual "gentle nudge" he took me by the forearms, looked straight into my eyes and said, "No, I mean it. I know you will do something really good with this and that's why I think you should have it."

He handed me the bracelet, and announced that he did not want to live in the same house where my mother and he hand lived until her death. He decided to move from Camarillo, California to Irvine (100 or so miles away from my sisters Judy and Cheryl—who had taken good care of both parents until that day).

I was happy, since I lived near Lynda and now I would at least be able to see my father on a regular basis. We played chess every Tuesday for more than two years. It has taken much more time than I expected for this project to be completed, because (as is my habit) when I took responsibility for completing a task my parents had given to me, I did much more than I originally expected because I did the best job possible, as I saw it. That was something I learned from both of my parents.

Anyway, the three of us went to see the movie *SELMA*, by Ava Duvernay. When we came out, into the lobby after watching that very powerful film (it brought a tear to my eye, I admit) I was astonished to see that my dad had tears rolling down his cheeks. He was crying! That was the first time I'd ever seen him cry! I was stunned a little, but I managed to ask, "What's going on?" or something like that.

He answered with the most remarkable part of his amazing story. He said, "My whole crew on Oahu was black."

At the time I didn't think much of it. I didn't know that history books had no mention of any black Marines at Pearl Harbor when the attack occurred. But, here was my father telling me that he was in charge a, "all black, no, maybe, I think they were all black...yeah, they were. I was in charge of an all black crew. They knew more about building than I did and I was the superior officer. I remember saying, 'I'm sorry I'm the superior officer. Just go ahead and tell me what you want me to order you to do.'"

Again, remember I did not know at this time that his comment was so important; so I asked him a follow up

By Gordon Richiusa

question from 14 years previous, "So in your 95 years of life, what is your greatest regret?" or, maybe "Do you have any regrets in your life?"

"I'm getting near the end," he answered quickly, as if he'd been thinking about this for a long time. "And I've been trying to rid myself of hate, and forgive even my enemies, but I will never forgive your grandfather Villani for how he treated your mom. I tried when she was alive, for her sake, but I just couldn't do it. But, my only real regret in my live is that I lost track of my best friend from high school, Floyd Fuji."

"You mean, when the war broke out?"

"Yes. I tried to find him a few times after the war, but I couldn't."

"Why? Did he end up in an internment camp? Manzanar?" I put two and two together in my head, suddenly and realized that maybe this was something my father had been trying to tell me for a long time, but that I'd just not picked up on the cues.

"Yes, I'm pretty sure that's the one," he answered. "I've always regretted losing touch with Floyd because we were such, good friends."

"So, why," I asked him in complete ignorance of the lesson he'd been trying to teach me for a long time, but that I was not sharp enough to comprehend, "Why, haven't you ever told me about your crew being black, or your best friend being Japanese American?"

"Because I didn't think ethnicity was important," he said.

By Gordon Richiusa

Part Two
Sam's Song

[Editorial Note: The following is a transcription of the video interview conducted by Gordon Richiusa of Salvatore Richiusa [his father] prior to the 60th Anniversary of Pearl Harbor. It is presented as it is recorded, with flaws and pauses, and sequence quirks.]

(Q) How well do you remember the morning of December 7, 1941?

"It was just before my twenty-second birthday. I was twenty-one years old. On the 11th, I turned 22."

(Q) How long had you been in the Marines?

"I joined the Marines in 1939…October 2nd. So, I was in a little over two years."

(Q) What was your expectation of war at that time?

"We expected that we would go to war, because England and Germany were fighting, and France was also involved in the war at that time. You never know what's going to happen."

(Q) *What was your rank?*
"At that time, I was a corporal."

(Q) *How vividly do you recall that particular day?*
"Well, I recall it pretty well. What do you want to know?"

(Q) *What time of day was it?*
"It was morning and we just got up, and we'd missed breakfast because we'd overslept. So, we were going to get some stuff at the Canteen in a place we called The Greasy Spoon out there. Junk food, we were going to eat that, and then we were going to play some cards. I think it was just before seven o'clock, Sunday, Sunday morning."

(Q) *Who were you with?*
"I can't remember all their names. A bunch of us were going to play some poker. A bunch of us were going to play, and I can't even remember all the names of the guys I was going to play poker with. There was Corporal Rudd, at that time, and Sergeant Sears…I can't think who they were now, but there was about two or three more guys who were going to play…five or six of us."

By Gordon Richiusa

(Q) *And all of a sudden you heard…?*
"Well, we heard bombing. We heard bombs going off, and so we went out. We were living in tents along the parade ground…the Marine barracks. And, so we just went outside the tents, looked up and saw planes flying around, and dropping bombs. We thought that was maneuvers of some kind. We saw the red circle on the side of the planes and we thought that was the red team probably, 'cause they had colored teams sometimes during maneuvers…and, I can't remember the sergeant's name that had been in China and he knew that was a Japanese…that those were Japanese planes. He knew the marking on there. And, he told us they were Japanese. They were not our planes. They looked quite a bit like one of our fighter planes that we had, but they were a little bit smaller. Then, suddenly they started flying over us! After they dropped the bombs in the harbor, they'd go over our barracks and started strafing us, shooting at us, and we were unarmed at that time. Anyway, they would fly over us and bomb Hickam Field, which was just on the other side of our parade ground there…at that time. Right now it's part of the civilian airport, the Honolulu Airport."

[Dad paused to check if this is what I wanted. I nodded to give him the go ahead and he continued.]

"When they started shootin' at us, we were told that that was a Japanese planes, then we were unarmed. All we had were rifles with no ammunition. The Sergeant…a few of the NCOs and sergeants had pistols, but they couldn't even fire those, because no one had any ammunition at that time. So, we went to the armory to get ammunition, and they told us we had to have a requisition for it."

(Q) Who was at the armory?
"I believe it was a lieutenant, a second lieutenant that was in there. Anyhow, he told us that we had to have a requisition and the Sergeant…"

(Q) Could he hear the bombing and see what was going on?
"I guess he could! We could hear it, but he wanted to refuse to give us any ammunition. So, Sergeant Sears…I think there were five of us who went in there…two privates, two corporals, and the sergeant. I don't remember the names of the privates but I do remember the names of the NCOs over there…and Sergeant says, 'Well, we're going to TAKE the ammunition.' [The Lieutenant] said we couldn't do that, but anyhow we just went in there and we took it. We just pushed our way in and loaded up some dollies that they had in there.

By Gordon Richiusa

We got ammunition, for rifle ammunition, pistol ammunition, and then machine gun ammunition for both 50 and 30 caliber machine guns…And, we also got machine guns, because we had none with us. We weren't assigned any machine guns for our outfit. So, we got some of those. We deployed among the heavy equipment there. I was in the Engineers. We had heavy equipment and we sort of used that for cover for us…bulldozers and carryalls and different kinds of tractors and stuff. We had all kinds of heavy equipment there. We were in among them, used those for cover and set up machine guns and [by then] everybody around us got ammunition for whatever they had and a…we started shootin'. The Marines [who] were deployed out there on the parade ground were credited with getting three planes that were flying over us."

(Q) How many planes were there altogether?
"Oh gosh, I never did know how many planes there were, but they'd come over us in waves, you know. There would be a little break and a bunch of 'em would start comin' [leading questions] over randomly it seemed like. They were flying very low. I would say less than 100 feet off the ground, and they'd fly over the barracks and were strafing us as they were going there. They just wanted to keep us down, the ones that were out in the open there. What they were trying to get were the ships

and the planes with their bombs. They didn't bomb us at all. They were trying to get the equipment there."

(Q) How long did this whole thing last?
"I think it was a little over an hour, because they came in two big waves. There must have been 50 to 100 planes and I'm sure some of them came over more than one time. Then, it stopped. There was no more bombing, but then we started preparing and everybody was armed by that time. Then we set up patrols. We set up machine guns on some of the trucks, patrolling the island…all around the island to make sure that there wasn't any landing coming in because we didn't know what was going to happen. Everyone was tense, because they wanted to be sure that we saw what was comin' then. There was no more firing until that night, and I'm not sure about the time this was…about an hour after sunset, hour and a half, maybe two hours, I don't know, after sunset we heard a plane, A plane I think. Everybody started shootin' at it. They picked it up on the lights. We DO know now that it wasn't a Japanese plane. It was an American plane. As far as I know we didn't hit it, but the place lit up like the fourth of July with tracer bullets just flying all over the place."

By Gordon Richiusa

(Q) So right after this hour where there was this continuous bombing and you guys are shootin' back a little bit…

"After the lull you mean? [After that lull] I don't know how many of us were on there, but we rode around in a truck—probably four or five of us—with a machine gun, a fifty caliber machine gun mounted on it and we were patrollin' around the island. When we got close to some pineapple fields, some of the natives there that were working in the pineapple field came over to our truck there and said they had killed a Japanese that had crashed his plane in one of the fields there."

(Q) Did you go check it out?
"No, we didn't go check it out. One of them gave me the knife he said he'd used to kill the [Japanese pilot]…that he said he'd used to kill him. I brought it home and gave it to my nephew Eddie. Last time I talked to him, he said he still had it."

(Q) How long were you in Hawaii after this?
"I left in March of '42 and we went back to the states. We went and did some training at Camp Eliot. We were forming a new engineering group."

(Q) That was the Sea Bees?

"No. The Sea Bees were Navy. We were Marine engineers. It was the second Marine engineers that I was in when then attacked us."

(Q) What was the new group called? Was there a name for it?
"The new group? I am not sure. We had so many different names. I think one of them was the 18th Marine Engineers. I can't recall all the different names we were under. We were still the same group, always mostly the same group of people that went in with the training…but we had new people that went in with us also, when we went back to the United States. Then, in September or October we went aboard ship. We didn't know where we were going. We went to a whole bunch of different islands [Umea?] and Samoa and stayed aboard ship all this time."

(Q) You didn't go ashore at all?
"When we were in [Umea] they didn't let us go ashore because they had some bubonic plague at that time. And, they didn't let us go ashore at Samoa. I don't recall what the reason was. I think they had some disease going around over there at that time, too. So, we didn't get to go ashore until a couple of weeks later after we left the United States…several weeks later. I don't know just exactly how long until we landed. We

spent one night in Australia and we got to go ashore there. The Australians met us there and invited us to go to the Y with them. They had some kind of a show they put on for us. Then, the next day we left. From there we went to New Zealand. We hit a great big storm, a storm where the waves were higher…I was aboard the Luriline, which was a pretty good size ship that they had commissioned to carry troops. It was a civilian ship company and was made for carrying troops after that, during all the war I guess. Cause' I don't think it was ever sunk."

(Q) So the waves were going over the top?
"Over the top of the big ship! If was very, very rough. They made us go below deck. The hatches were all closed, I guess so they wouldn't get swamped. The air got stale, but we tried to play cards while this is all happening. Sittin' on a blanket…we would slide all the way across the floor…the ship was tilting and we'd slide clear across the rooms there. Anyhow, several days later we got to New Zealand. I was in New Zealand for pretty close to a year. We built a camp at one of the Maori villages there. I can't recall what they call that area. We built a camp that was supposed to be for people coming back from over seas…to go out there and rest and train and whatever they were doing. It was not for Americans. I believe it was for the New Zealand troops that were coming back from Africa. They

were fighting in Africa. Anyhow, when we finished that, we were there almost a year…then we went to Hawaii."

(Q) So this was 1942, 43?
"This was late in '43 we went to Hawaii. We may have been there OVER a year. Seems to me we might have spent two Christmases there. We went to the Big Island and we built a camp over there for people coming back for recreation and things. We built a theater there at this camp. I don't recall the name of this camp, but I have that information and can give it to you if you want it."

(Q) So, when did you ship out again to…
"Like a said, we went to Hawaii and we stayed on the Big Island for several months. We finished our camp there they put us all aboard ship…"

[Archival video recording jumps OUT OF SEQUENCE…]

"There were ships all over the place over there."

(Q) And you saw them getting hit?
"I didn't see them getting hit, because the Marine barracks was between us and the harbor. But, we visited after the attack

we visited a lot of places over there and we could see some of the ship... A couple of ships you could see were gone aground, and also, this is after the attack, several days later they brought up...I think they said it was a two-man submarine. It was a very small submarine...I would say thirty feet long at the most, very small, a Japanese [sub] that they'd captured... We saw a couple of planes that had gone down and a piece of one plane and I made mom a bracelet...I don't know if you ever saw it...it was a double heart bracelet."

(G) "I think I have."

"And, I have...I still have a piece of that plane in one of those albums I made. It was made out of wood. I have a piece of a plane in there, and that's where I have my commendation..."

(Q) Oh, from Roosevelt?

"I'm pretty sure it was from Roosevelt...It must have been...at the time."

(G) "Yeah."

[Recording JUMPS BACK TO STORY AFTER THE ATTACK]

HEROES' HEARTS ®

(Q) So, what happened next? What happened after December 7th?

"We went aboard ship and held maneuvers on Maui. Then, a…we got orders to go to Saipan. We didn't know we were going to Saipan, but we got orders while we were aboard ship that we were going to continue to battle, going to an island. We knew we were going to an island but we didn't know where. While we were aboard ship, we were attacked by Japanese' planes. This was a week or ten days after we left Hawaii, four or five days before we landed. We landed in Guam…I'm sorry…we didn't land in Guam. We landed in Saipan, which is in the Marianas. So is Guam. Other troops landed in Guam. We took that island. We dug in that first night. There was an airfield and we were between the airfield and the ocean. Late that night, probably around midnight, I'm not sure of the time. The Japanese had a banzai attack on us. We had gone across there, but after we took the beachhead there, we stopped at the airstrip; other Marines had come in behind us. They went on and had taken the airfield and had gone across the airfield. During that night, the Banzai attack happened. By that time, we had gun in placements set up all along the airfield. When they were coming across the airfield, we were just shooting 'em down, and they just kept on coming. Up to the north of us, a town called [Gairpan?] they overran that town. They overran the Marines up there and they took part of the artillery. I believe it was the 10th Marines. They overran them. I not sure how, but they did get a lot of our equipment. Then, by the next day we took all that part back. The Banzai, when they got over halfway across the airfield and I guess they gave up. They retreated…they retreated after that, back to the other side of the airfield…"

(Q) What made them different?
"The next morning…What made them different was that they were screamin' and hollerin' and they were out in the open, because they had no cover, just right out in the open. And, a…I guess they were trying to scare us with their hollerin' and stuff, but we were just cuttin' 'em down. And, the next day we were pretty complacent. We had captured, over to the side of the airfield there was a Japanese ammunition dump. Maybe seventy-five or a hundred yards north east of us. We had gone through there and made sure there was no enemy in there. It was all loaded, all full of all different kinds of Japanese equipment and ammunition. Sometime on the second day they started using their artillery and aimed it at that ammunition dump they had there…and they exploded it."

(Q) Blew up their own ammunition?

"They blew up their own ammunition. By doing that they had fragments of anti aircraft ammunition and bombs were exploded. Some pieces of [shrapnel] were about 18 inches long, in fragments, and they'd come flying through the air. And, one piece landed in the foxhole I was in and killed a guy right next to me…[cut his head off]…and, a…I don't even recall his name."

[NOTE: My father's voice was full of pain here, and this and other personal observation is why we are hoping to help PTSD survivors.]

(Q) Were you injured then?

"No, I was not injured."

(Q) When were you injured?

"The only day I was injured at all, and I did not report it, was on the first day at Pearl Harbor. I got hit in the ankle with a dying bullet. I just burned my flesh, and it just sort of tore a little bit of the flesh away there. That's the only time I was injured during the war. And, I didn't report it because I didn't wanna…I thought we were going from there on toward Japan, you know, and attack them. But, they sent us back to the States."

By Gordon Richiusa

(Q) So, you were angry at the Japanese…
"I was angry at the Japanese. I didn't want to have to say that I was hurt. It [the wound] got infected a little bit, but it went away after a couple of weeks."

(Q) Where do you think you picked up malaria?
"I picked up what they thought was malaria, at that time…and I believe it was the onset of dengue fever, instead of malaria…it was on Saipan. We'd been there probably a week or two. I was getting the chills and they thought it was malaria. Maybe it was because I had recurring chills and they told me that I had dengue fever, from the chills and fever that I had first. Then, I started aching…my whole body was aching. That's how they discovered the dengue fever, but after the war…the fact is, while I was still in the service, several times I got recurring chills and they told me that I had malaria. So, I guess that's where I got the malaria, in Saipan, but also dengue fever. Anyhow, I lost probably twenty pounds or more, maybe even thirty pounds from this dengue fever…well, from the dengue fever. It took me a week, ten days to get over it. Probably the result of the dengue fever also from getting sick and tired of eatin' sea rations…When I got better they had us goin' out on what they call clean up. We had a group goin'out and through the caves. Some days we'd capture several

Japanese who were holed up. Some of them were sick with dengue fever and probably malaria. I'm not sure what they had. We'd also capture a lot of the Chammaris {Chomorros?} the native people of the Marianas. We'd try to segregate the Japanese from the Maris. The Japanese we had to keep more control over 'em. The Chamarris wanted to get out so we'd use them to help us pick up the dead bodies and stuff. …During that time I picked up a lot of Japanese rifles, and Japanese souvenirs, but I was more interested in eating. Ever chance I'd get, I'd go aboard ship, aboard the ships that were coming into the harbor there and trade all these rifles and helmets and whatever we had, {pronunciation is Sam Are ees} Samurais. We had a couple of Samurais and I wished I had known that…they said that the handles…they had jewels under the wrappings on 'em, under the wrapping in the cord…"

(Q) In the swords?
"Yeah, and in the handles they said that a lot of them were encrusted with jewels. Anyhow, I probably would have traded them off for food…I got food from the ships, for those. I used to get the rice…"

(Q) A good sandwich is worth its weight in jewels, huh?

By Gordon Richiusa

"What we got actually, powdered milk, chocolate and so used to make what we called…I don't know actually what it was…but we called it rice pudding. We'd use the Japanese rice that we'd get out of the caves, boil all that and put chocolate and powdered milk in it, and sugar to sweeten it, you know. It was pretty good.We lived on that for a while. After that…I don't remember how long…After we secured the island…then, we went and took the island of Tinian, which was a neighboring island to Saipan. The battle didn't last a couple of days, the part that I was involved in, but we still had the clean up. We'd capture some Japanese and some…I would say, almost every day one or two Marines would get killed…It was worse there, in Tinian then it was on Saipan, at least for our outfit it was."

(Q) How would they get killed?
"The snipers would start shooting at 'em from different places, you know, and sometimes we could find out where they were shooting from and sometimes we couldn't. And, sometimes we could capture or kill 'em, you know? But, anyhow, it kind of affected me mentally, this did. After you had so many months overseas, you got points. They would rotate you OUT, and send you back home…but, when I had enough points, well they kept me out there because of my

rating. They said they held me there for the convenience of the Government. Even if I had enough points to go…I had more than enough, almost double the points needed…finally, they let us go home, from Tinian. We went home. I'm not sure what the date was there."

(Q) How long were you home?
"We got home…I think it was March of '45. They kept us there and we had pretty good duty. They let us go on leave a lot. Just before the end of the war…In fact, the day that the war ended, they were sending us back overseas. They'd just told us that we were going to go back overseas. We had gotten…we'd packed our sea bags and they had been sent too the dock…and the war ended. We didn't all get to stay. Everybody who had enough points got to stay. So, I got out. The war ended, I think early September of 1945…they kept me there until the 27th, and finally they discharged me out. And, that was it."

(Q) I wanted to ask you about when you went to the 50 year reunion, but do you want to take a break?

{Sam goes to the restroom}
[Recording OUT OF SEQUENCE]

By Gordon Richiusa

(Q) Do you have any good memory whatsoever from this period of time?

"They sent me back to Camp Pendleton to await discharge…When I got discharge, of course that was a very happy time too."

(Q) Any other incidents or anything that you think would be interesting to others or that you wanted to talk about? Or, anything that I'd forgot to mention or ask you about?

"I don't know…Going through the caves we captured groups and one time, we went through the caves and didn't see anybody. Then we went on down and when we were coming back they got a group of about 10-15 Japanese out of this one cave that we'd already entered. We didn't go on back…cause they had supplies, when you'd go inside, there would be supplies piled up so we didn't climb over all that stuff. But, we would throw a couple of grenades back there, and you know, past where we could see. We didn't hear anything back there so we continued on out. Evidently we shook them up or something 'cause they came out later on, holding their hands up…Ten or fifteen of 'em, a group of 'em anyhow. The most that we'd captured before was one or two guys at a time. We helped this other group take 'em on down. I was happy that they didn't attack us when we were in there. We had gone in there and got a bag of rice, about fifty pounds of rice we'd take that back with us...Oh, I don't know. There's so many things and my memory has some things out of context, too. I'm not sure if [it] happened before this or after this. I know that they happened."

(Q) Who would you say were your best friends?
"While I was in the service? Most of the group who went to New Zealand with, they got to go home because they were not kept for the good of the service like I was kept there. My best friends at the end of the war...one, let's see, ohhhhh, Abbott...I can't remember what his first name was. He was a corporal...one of my best friends. I don't even know if I ever knew his first name because we just called each other by the last names."

(Q) Your rank and your last name...
"Most of the guys called me Rick, instead of Richiusa. Let's see, Abbott was probably my best friend and Walker...and Ray...I knew his first name. Now, I can't think was his last name was...Anyhow, he lived in Iowa...He went back to Iowa...and Walker lived in Iowa, and part of the time there was a war correspondent named Larry Porter. He was one of my good friends. Abbott, right after the war he went back

home…he got married right after we got back from Tinian. Him and his wife went back to Provo, Utah. I haven't heard from any of them since. I've had no contact with any of the guys at all. I never saw them over at Pearl Harbor at the 50th Anniversary."

(Q) You know about the Pearl Harbor movie coming out this summer?
"Yeah, I know."

(Q) Are you looking forward to seeing it?
"I'd like to see what they did with it, see how authentic it is…I'm curious. Any other questions you can think of? I have some stuff at home…I think I still have it…I have the commendation…from President Roosevelt."

(Q) This was right after the attack at Pearl Harbor. I thought there was another commendation from the 50th anniversary…
"There was no commendation there, they just had this parade for us. They saluted us, you know. They really treated us well. Everybody treated us well... the services…Also, we went to the Arizona. We saw that."

(Q) How'd that make you feel?
"It was kind of eerie in a way knowing a ship was under there, right where we were there. You could see bubbles coming up. I don't know if you've ever been there or not."

(G) No I haven't
"It's kind of interesting. I say bubbles, but it was fuel…fuel oil still coming up after all that time. I don't know if it still is, now, but at that time it was still coming up."

(Q) Could you see the Arizona from the base? Or, was that too far away?
"You mean before the attack?"

(Q) So, here we're back and 50 years later you are invited by who to go to…
"My brother-in-law Ralph asked me if I'd like to go back. And, I said yeah, but I can't afford it. So, he got tickets and sent us, got a hotel and transportation back and the first day we got there, before the 7th, several days before the ceremony…at night, one of the Marine Colonels invited us over to…first, before that they had a ceremony for us, after the parade. We marched down the street in Honolulu. Then, we toured around the island and that evening they invited all the survivors, they

invited us all to, especially the Marines, to go to a ceremony they had right at the end of the parade ground. They had bleachers there…"

(Q) Same area where you guys were?

"Same area where we had our tents. There were no longer any tents there. The barracks were still there. The reason we had to go in tents was because the barracks were all full of other people. They had gone in there before we got there. And, besides that, once we had got up…when we had built up to a certain point, when we built that camp, we moved up there. The buildings that were completed were our barracks, and that's where we stayed. That camp was about halfway between Pearl Harbor Naval Base and downtown Honolulu. It was above a big cane field. We did visit…but I'll go back, the first day we got there, the first day we arrived at the hotel, the first person I met was somebody who was in my outfit….Mom and I met this couple up there and he was a Marine. He was the first person I met and the first person we talked to when we arrived at the hotel, and he happened to have been in my outfit. I never met anyone else from my outfit."

(Q) Did you remember him?

"I did NOT remember him. He worked in the headquarters and I worked in the sawmill…the construction part of the engineers."

(Q) What about this other friend you have…in Oregon or something? Somebody that you knew from there…some sort of nickname you gave him…
"You're talking about Gerald Warner…Dodo? He wasn't in the service with me. He was just a friend. We went through grade school and high school together. He wasn't in the service with me….After we got back, I never met any of those guys again. One of them lived over by Redondo Beach…But, this guy we met at the hotel…We hit it off and…Mom and I and him and his wife we went all over the island besides."

(Q) Wasn't there some kind of mix up with the hotel room?
"Yes, that's right. That happened when we got there at the hotel…The Japanese…When reservations had been made, the Japanese did NOT own the hotel. But, when we got there, the Japanese had bought out the hotel. I don't recall which hotel it was; it was a nice hotel. They told us that they had to do some construction work over there, but we know better. The whole thing was all sold out to Japanese tourists that came there. They gave them preference over all the servicemen. We were

affected, but I don't know whether ALL the service men were affected, but they put us in some little old flea bitten, old beat up motel…You entered the rooms from the outside. The office was downstairs and you'd go up on the outside, up some stairs. They put us up in this seedy old room. The place was full of bugs and other creatures, geckoes and stuff. We'd only sleep there at night then we were gone all the time…anyway, to get back…after that service that they had for us at the Marine barracks, parade ground, we were all invited to a colonel's…all the WWII veterans for this big luau. All the food you could imagine, and drink…all the drinks you wanted. They really treated us well. And, while we were there we could go to the Canteen and could buy anything we wanted and no taxes for servicemen, and we got the same treatment as the servicemen…everything you wanted…tax free…I didn't smoke, but cigarettes were tax free…at that time, I had already quit smoking. We were really treated well…very well, by the military and by the people there except for the hotel…the hotel group. My brother-in-law got reimbursed for the hotel. I put up a stink about it…except for that, it was a very nice trip."

(Q) You were on other islands that Saipan and Tinian…
  "In battle, Saipan and Tinian were the only battles. We went to a lot of other places."

(Q) If you had to rank 'em what would have been the worst experience? The most frightening, the most...

"My first landing was frightening. The attack on Pearl Harbor I was not afraid. I was so mad about it, that I was not afraid. But, our first landing you got to think about it. You know, when it happens to you suddenly you have a different reaction, but when you think about something, a landing, going in…in landing craft…they'd be shootin' at you and you just continue on going in, you know? Thinking about it before, even, they started shootin' at us I was scared to death at that first landing. I had stomach cramps and everything else! Once you get to fighting, then you don't even think about being afraid."

(Q) If you had to pick something to be your best memory of that period between the day you were bombed in Pearl Harbor and the day they said 'The War's Over'?
"When I got to go home, from Pearl Harbor, I got married about a month and a half when we got back to the states. That was a very good memory. Then, when I came home from Tinian, after we left Tinian we came back to The States and I got to see my oldest daughter for the first time. She was just

two years old, just a little older than two years old. I had not seen her. She was born when I was in New Zealand."

(Q) The day that Japan was bombed with the A bomb, do you remember that?
"Yes. I was at Camp Pendleton at that time. Of course we all cheered that we had it. In fact, before that we had gone in and bombed it…some of the planes didn't make it, but we cheered that too. We were looking forward to the war ending…there was so much damage done in Japan, from what we were hearing we were quite sure the war was just about over, but they were still making plans to send us overseas. I was still kind of worried. But, we were happy. That was one of my happier times probably when they told me I didn't have to go aboard ship."

HEROES' HEARTS ®

[NOTE: Archival video recording abruptly ends here. No notes or other recordings have been located. However, researchers and historians, including Daniel Martinez, Park Ranger and official historian at the Arizona Memorial and Heroes of the Pacific Museum at Pearl Harbor. A photograph of Gordon Richiusa and documentary film producer Michelle Manu meeting Daniel at the Arizona Memorial and showing him the original Heroes' Hearts bracelet is in the following photo gallery.]

By Gordon Richiusa

Photo Gallery

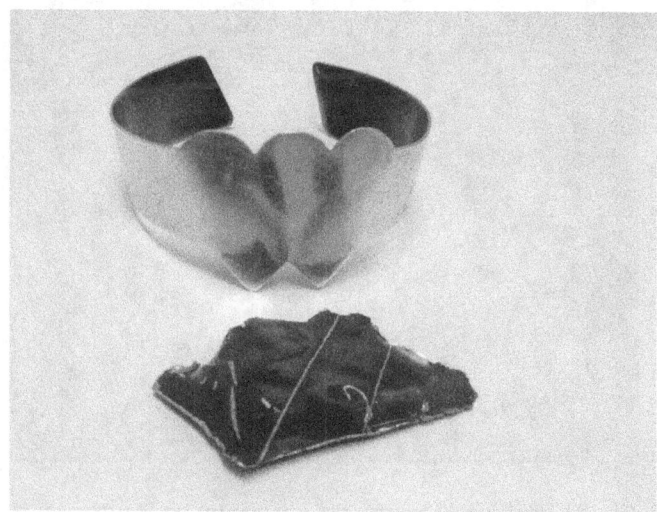

Two Hearts Beat as One Original and the "unmodified" piece that helped us identify the exact Japanese plane.

Floyd Fujii at Manzanar, courtesy of Japanese American Historical Museum.

By Gordon Richiusa

Sam and Mae on flight to Italy for 56th Wedding Anniversary. Lynda and Judy accompanied them.

HEROES' HEARTS ®

Mae "willingly" helps Sam with a project. Then, Sam takes the credit.

By Gordon Richiusa

Gramma and Grandpa Richiusa family, with young Sam lower right. Below, these children multiplied into a large brood.

HEROES' HEARTS ®

By Gordon Richiusa

Sam with Villani "brother in laws" from left: Red, Sam, Ralph, Ray. Red is holding his daughter (my Villani cousin) Paula.

When Sam returned home and met his 2 ½ year old daughter for the first time...Lynda Lee (Photo Editor for this book).

By Gordon Richiusa

Sam showing where he was stationed during the attack on Pearl Harbor.

Rare photo, taken by Sam Richiusa, of the Military Battle of the Bands, December 6, 1941. The Japanese attack was likely scheduled for Sunday the 7th, in part because it was well known that the U.S. Military was enjoying themself on Saturday.

By Gordon Richiusa

Richiusa and Villani families at joint summit, family gathering. Ralph Villani recalled that the two families were together on December 7, 1941, even though Sam and Mae were yet to be married.

HEROES' HEARTS ®

Sam and friends toured Oahu and recorded adventure, the day BEFORE Pearl Harbor was attacked.

By Gordon Richiusa

Mae and Sam asked their son, Gordon (who had, on a writing assignment become a minister in the First Church of God the Father, to perform 50th Anniversary re-marriage vows.

Rare photo, presumed to have been taken by Sam Richiusa, of the Barracks he and his crew built, dated December 6, 1941.

By Gordon Richiusa

The first "perfect replica" laser copy made by students and faculty of Saddleback College's VETS program.

Sam received (and kept in his logbook until the day he died) a commendation for conceiving this method of raising tents at Hickam Field.

By Gordon Richiusa

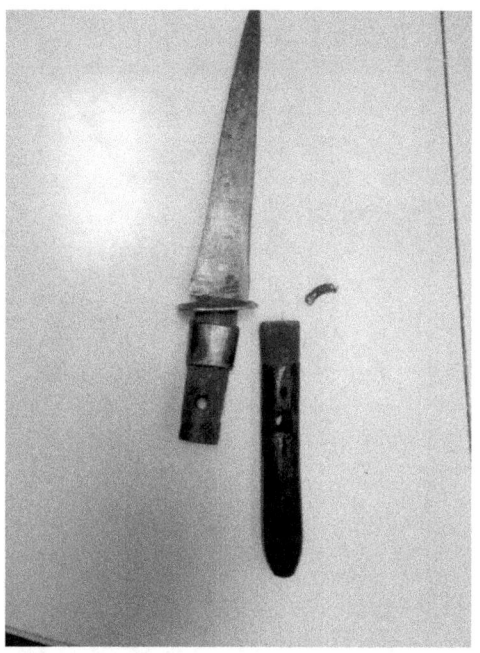

Could this be the knife given to Sam by Hawaiian Native? It was donated by Dana Moffat, the daughter of my cousin Eddie Shane.

HEROES' HEARTS ®

THE WHITE HOUSE
WASHINGTON

July 15, 2016

Mr. Gordon Richiusa
Laguna Woods, California

Dear Gordon:

This is just a quick note to thank you for sharing your father's story.

Your father is among a generation of heroes who, amid the harshest of circumstances, helped free a continent from the grasp of tyranny and change the course of an entire century. America will be forever grateful for the sacrifices made by our World War II veterans, and in their extraordinary example we see the true meaning of patriotism—a devotion so deep they were brave enough to risk their lives to protect the ideals for which our Nation stands.

Each day, I am thankful for the service of all our troops, our veterans, and their families. I trust you take pride in your father's remarkable legacy, and I wish you all the best.

Sincerely,

[signed] Barack Obama

Letter from President Obama, regarding Sam's story that his crew at Pearl was black and best friend Japanese American.

By Gordon Richiusa

At 75th Anniversary of Pearl Harbor, on Oahu, (L) the flagpole where exact replica was placed. (R) Kumu Michelle Manu solemnly dances on Hickam Field beach, wearing original Heroes' Hearts and Ed Hoffman copper bracelets.

Oldest daughter Lynda (L) and second child, Judy (R) at 75th Anniversary Event, Hickam Field, Oahu, December 7, 2016 wearing original and replica bracelets.

By Gordon Richiusa

The original Heroes' Hearts, Two Hearts Beating As One, displayed with three hand-made copper versions by Auschwitz Survivor, Ed Hoffman. Ed asked to make bracelets of copper because he said, "Copper is healing."

HEROES' HEARTS ®

A commemorative brick that is placed near a flagpole at Saddleback College's Veterans Memorial.

2nd brick that Sam and Mae displayed proudly at their home in Camarillo, CA until Mae died and Sam moved to Irvine.

By Gordon Richiusa

Sam's log album where rare photos and the "unmodified" fragment was kept for 75 years.

HEROES' HEARTS ®

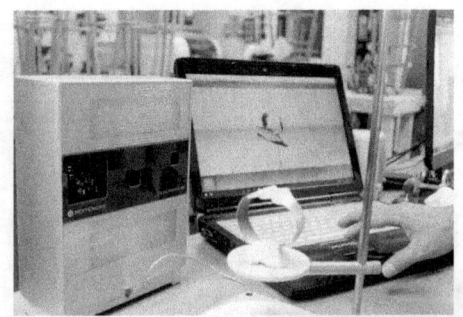

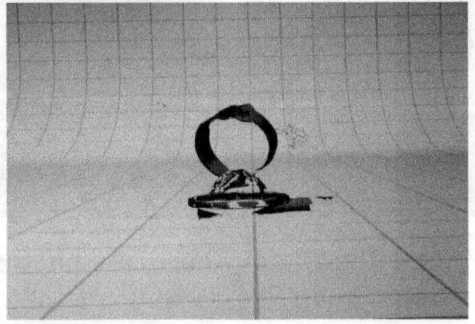

Hand of Glen Stevenson, VETS instructor at Saddleback, and computer image from first laser scan of original bracelet.

By Gordon Richiusa

Mae and Sam at their first and future home site in Sunland, CA

HEROES' HEARTS ®

Two beaches on Oahu, photos taken by Sam Richiusa, dated December 6, 1941. (R) May be Hickam Beach.

By Gordon Richiusa

Dated December 6, 1941, from Sam's logbook, the Engineers' motor pool at Pearl Harbor.

HEROES' HEARTS ®

Children with proud parents, from Left: Judy, Cheryl, Gordon, Mae, Gary, Sam, Lynda.

By Gordon Richiusa

Michelle Manu, Daniel Martinez, Gordon Richiusa at Arizona Memorial, December 6, 2016.

HEROES' HEARTS ®

Sam on grass in front of Royal Hawaiian dated
December 6, 1941.

By Gordon Richiusa

Sam refereeing mock sword battle and with Boggs and Winslow December 6, 1941

HEROES' HEARTS ®

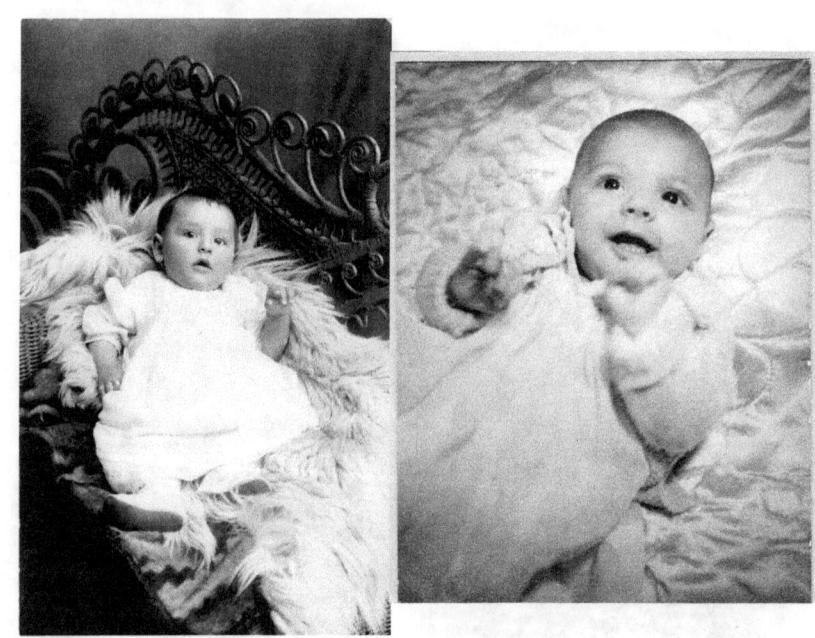

Salvatore (Sam) left and Philomena (Mae) right posing for their baptism photos.

By Gordon Richiusa

Do Something Good

The Legacy of Heroes' Hearts®

It was not until after both my mother and father had passed away that I started to understand the incredible story that surrounded the Two Hearts Beating As One© bracelet that my father made for my mother, or to try and honor my dad's wishes that I do something good with the bangle he'd given to me at my mom's memorial service.

I think both of my parents would be happy with the result of my efforts. With the help of my siblings, (oldest to youngest) Lynda, Judy, Gary and Cheryl, my son Travis and many others we've started a non-profit corporation called Heroes' Hearts Inc. with the motto: Do Something Good with the mission of opposing prejudice and discrimination with projects (including the exact replica bracelets and Heroes' Hearts® brand) to benefit those who've experienced trauma, such as PTSD. We believe that every person who puts others in front of themself is a hero, and we promise to continue to partner with veterans groups, businesses, and other

HEROES' HEARTS ®

individuals to honor any and all expressions of selfless love…for family, friends, or country.

Please visit www.HeroesHeartsBracelet.org for more information about our projects, to make a tax-deductible donation, or purchase trademarked merchandise including the exact replica bracelet. All proceeds will enable us to continue achieving our goal as expressed by my father…Do Something Good.

Copyright and trademarked photos available through:
www.lyndaleephoto.com

By Gordon Richiusa

HEROES' HEARTS ®

www.ingramcontent.com/pod-product-compliance
Lightning Source LLC
Chambersburg PA
CBHW060518300426
44112CB00017B/2716